ASTROFICTION

A Worldbuilder's Reference for
Science Fiction and Fantasy

By Ingrid Moon

Contents

Introduction

There is no right or wrong way to create a universe of our own.

However, there are some basic rules that people are accustomed to, such as gravity. There are also specific terms we use to describe the phenomena of the universe around us. You can, of course, make up your own, or create new laws of physics, but chances are, you'll want to refrain from creating too much suspension of disbelief.

This book is a reference to help you find and use those terms, and to form a basis for any rules you plan to break, or any new systems or magic that you plan to create. It is for people who want their worldbuilding to make sense, not just as it fits into the story, but in the subconscious of their readers, players, and viewers.

I hope you will get the most out of it, and I look forward to seeing what you create with it.

~
Ingrid

Introduction to Galaxies

Galaxies are listed in order of commonality, from trillions known (Dwarf) to a few dozen known (Ring, Jellyfish).

- **Dwarf Galaxies**: Dwarf galaxies are small in size and typically have lower luminosity compared to larger galaxies. They come in various forms, including dwarf ellipticals and dwarf irregulars.

- **Elliptical Galaxies**: Elliptical galaxies have a more spherical or ellipsoidal shape, without distinct spiral arms. They vary in size from small to massive and are often composed of older stars.

- **Irregular Galaxies**: Irregular galaxies have no well-defined shape and can appear chaotic. They often contain young stars, dust, and gas. The Large Magellanic Cloud is an example.

- **Spiral Galaxies**: Spiral galaxies are characterized by spiral arms that wrap around a central nucleus. Examples include the Milky Way and the Andromeda Galaxy.

- **Lenticular Galaxies**: Lenticular galaxies have a disk-like structure similar to spirals but lack prominent spiral arms. They often have a central bulge and a disk.

- **Barred Spiral Galaxies**: These galaxies have a central bar-shaped region that extends outward from the nucleus. The Milky Way is considered a barred spiral galaxy.

- **Compact Galaxies**: Compact galaxies are small, densely packed systems with a high concentration of stars. They are often found in galaxy clusters.

- **Blue Compact Dwarfs**: These galaxies are small, blue, and undergoing intense star formation. They are known for their high rate of star birth.

- **Low Surface Brightness Galaxies** (LSBGs): LSBGs have a low luminosity per unit area, making them challenging to observe because of their faintness.

- **Starburst Galaxies**: Starburst galaxies undergo a burst of intense star formation, resulting in a high rate of new star formation and often appearing luminous.

- **Ultra-Diffuse Galaxies** (UDGs): UDGs are faint galaxies with low surface brightness, making them challenging to detect. They can be both dwarf and giant galaxies.

- **Quasar Host Galaxies**: Quasars are extremely bright and distant celestial objects powered by active galactic nuclei. They are often associated with host galaxies.

- **Peculiar Galaxies**: Peculiar galaxies exhibit unusual characteristics, such as distorted shapes, tidal tails, or intense starburst activity. The Antennae Galaxies are an example.

- **Hybrid Galaxies**: Hybrid galaxies exhibit characteristics of more than one primary galaxy type, often resulting from complex interactions or mergers.

- **Cannibalistic Galaxies** (Galaxy Cannibals): Some galaxies, known as cannibalistic galaxies, have absorbed smaller galaxies through mergers. These interactions can result in unique structures.

- **Seyfert Galaxies**: Seyfert galaxies are characterized by an active galactic nucleus with a supermassive black hole that produces intense radiation, including X-rays.

- **Radio Galaxies**: These galaxies emit powerful radio waves and often have active galactic nuclei with supermassive black holes.

- **Green Pea Galaxies**: Green pea galaxies are compact, small galaxies with intense star-forming regions that emit green light due to strong oxygen emission lines.

- **Ring Galaxies**: Ring galaxies have a ring-like structure surrounding a central core. They often result from collisions or interactions with other galaxies.

- **Jellyfish Galaxies**: Jellyfish galaxies have long tails of gas and stars, resembling the trailing tentacles of a jellyfish. These tails are a result of galactic ram-pressure stripping in galaxy clusters.

Oversimplified Milky Way Galaxy

Our galaxy is approximately
100,000 light years across

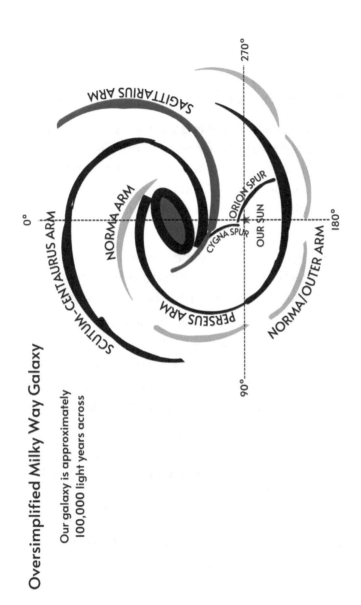

A Primer on Gravity

Gravity is a fundamental force in the universe that acts between all objects with mass or energy.

- **Mass attracts mass.**

- Important fact: **There is gravity in space.** It is just very small when it is far away from objects with mass or high energy. It is called **microgravity**, and while it may approach zero, it can never be zero. If it were zero, planets would not orbit their stars, the moon would not orbit the Earth, etc.

- However, once outside the "**gravity well**" of a celestial body (such as a planet), it becomes microgravity and objects will become weightless.

- On the ISS, or in Low Earth Orbit (LEO), astronauts experience "weightlessness" or microgravity because they are **in a state of free fall** around the Earth. At that distance, around 420 kilometers, Earth's gravity is **90%** of gravity on the surface.

- Astronauts experience true weightlessness when they are beyond the Earth's gravitational influence, approximately **384,400 kilometers** (about 238,855 miles) from Earth's surface.

- Massive celestial bodies like planets, stars, and galaxies have **significant mass** and therefore exert strong gravitational forces.

- Gravity follows an **inverse square law**, which means that the force of gravity weakens with distance squared. If you double the distance between two objects (x2), the gravitational force weakens by a factor of four (2^2).

- Kepler's laws of **planetary motion** describe how objects in orbit around massive bodies, like planets around the sun, move under the influence of gravity.

- The gravity of a planet depends on its mass and density. (see Planets & Worlds below.)

- **Gravitational waves** are ripples in spacetime caused by the acceleration of massive objects. As such, **massive objects warp the fabric** of spacetime around them. Objects in motion follow **curved paths**, which we perceive as the force of gravity.

- Gravity plays an important role in **the formation of cosmic structures**, including galaxies, star clusters, and large-scale cosmic filaments.

- The majority of the universe's mass is believed to be made up **of dark matter**, which does not interact with light but exerts gravitational influence.

The Scale of the Universe

The following table describes the scale of the universe in terms of orders of magnitude.

"Orders of magnitude" is a way to measure the size, scale, or quantity of something by expressing it as a power of 10. For example, going from 1 to 10 is a change of one order of magnitude, and going from 1 to 100 is a change of two orders of magnitude. It's a way to simplify and understand the relative size or scale of things.

* Human height is about 1-1.5 meters. Rounding that to order of magnitude = 10^0 = 1

Object	Diameter/Distance/Scale
Observable Universe	10^{26} meters
Observable Universe (light-years)	10^{10} light-years
Superclusters of Galaxies	10^{24} meters
Galaxy Clusters	10^{23} meters
Galaxies (e.g., Milky Way)	10^{21} meters
Solar System	10^{13} meters
Planets (e.g., Earth)	10^{7} meters
Cities (e.g., New York City)	10^{5} meters – for reference 100,000m
Human Height	**10^{0} meters ***
Cells	10^{-5} meters (micrometers)
Molecules (e.g., DNA)	10^{-9} meters (nanometers)
Atoms (e.g., Hydrogen)	10^{-10} meters (angstroms)
Atomic Nucleus	10^{-15} meters (femtometers)
Subatomic Particles (e.g., Protons)	10^{-18} meters (atmometers)
Quantum Scale (Planck Length)	10^{-35} meters

Galaxy Dynamics

Astronomers use various techniques, including telescopes and simulations, to study the dynamics and evolution of galaxies.

- **Formation of Galaxies:**
  - Galaxies are thought to form from the gravitational collapse of large clouds of gas and dust in the early universe.
  - Tiny density fluctuations in the early universe gave rise to slight over-densities, which became the seeds for galaxy formation.

- **Star Formation:**
  - Within galaxies, stars form from the gravitational collapse of dense regions within molecular clouds.
  - Star formation is a continuous process, with some regions of galaxies being active nurseries for new stars.

- **Spiral Structure:**
  - Spiral galaxies rotate, with stars and gas moving in nearly circular orbits around the galactic center.
  - Spiral galaxies often have distinctive spiral arms that are not rigid structures but patterns of star formation and movement.

- **Elliptical Galaxies:**
  - Elliptical galaxies have more random motion and lack the well-defined spiral arms of spirals.
  - They often form from the collision and merging of smaller galaxies.

- **Galactic Collisions:**
  - Galaxies can collide and merge due to gravitational interactions.
  - Collisions can trigger intense bursts of star formation and reshape galaxy structures.

- **Interactions and Tidal Forces:**
  - As galaxies approach each other, gravitational tidal forces can distort their shapes.
  - Tidal interactions can also lead to the formation of long tails of stars and gas in merging galaxies.

- **Supermassive Black Holes:**
  - Many galaxies host supermassive black holes at their centers, including the Milky Way.
  - These black holes can influence the motion of stars and gas in their vicinity.

- **Galactic Cannibalism:**
  - Larger galaxies can "eat" smaller ones through mergers and gravitational interactions, growing in size over time.

- **Stellar Populations:**
  - o Galaxies can have diverse populations of stars, with varying ages, compositions, and lifetimes.

- **Galactic Clusters:**
  - o Galaxies are often found in groups or clusters, bound by gravity.
  - o These clusters can contain hundreds to thousands of galaxies.

- **Dark Matter Halo:**
  - o Most galaxies are believed to be surrounded by massive dark matter halos that provide additional gravitational influence.

- **Galactic Evolution:**
  - o Over billions of years, galaxies evolve, changing in size, shape, and composition due to internal and external interactions.

Not listed by commonality:

- **Irregular Galaxies:**
  - o These galaxies are less structured and less symmetric in appearance.
  - o LMC and SMC are irregular galaxies.

Galaxy Groups

Our galaxy exists in a cluster of galaxies called a **local group**. Most galaxies exist in groups.

- All the galaxies in the Milky Way system are within about 250,000 lightyears of one another, contained within a spherical cloud of dark matter.

Our local group includes:

- Andromeda Galaxy (M31) – a spiral galaxy slightly larger than the Milky Way

- Triangulum Glaaxy (M33) – a smaller spiral

- Large Magellanic Cloud (LMC) - irregular

- Small Magellanic Cloud (SMC) - irregular

- M32 and M110 (satellite galaxies of Andromeda) – compact elliptical and dwarf elliptical respectively

- And many more

Black Holes

Black holes are fascinating and enigmatic objects that challenge our understanding of physics. While many facts about them have been established through observation and theory, there are still hypothetical aspects and unresolved questions that continue to intrigue scientists.

- **Types of Black Holes:**

    o <u>Stellar-Mass Black Holes</u>: Formed from the collapse of massive stars, they typically have masses between a few and tens of times that of the Sun.

    o <u>Intermediate-Mass Black Holes</u>: The existence of black holes in the range of hundreds to thousands of solar masses is still debated.

    o <u>Supermassive Black Holes</u>: These are found at the centers of galaxies and can have masses ranging from millions to billions of times that of the Sun.

- **Size and Mass**: The size of a black hole is determined by its mass and is characterized by the Schwarzschild radius, which is the radius of the event horizon. The formula for the Schwarzschild radius is proportional to the mass of the black hole.

- **Formation**: Black holes form from the remnants of massive stars that have undergone gravitational collapse after exhausting their nuclear fuel.

- **Singularity**: At the center of a black hole, there is a singularity—a point of infinite density where gravity is infinitely strong. Classical physics breaks down at this point.

- **Event Horizon**: The boundary surrounding a black hole is called the event horizon. Once an object crosses this boundary, it can never escape, not even light. It marks the point of no return.

- **Hawking Radiation (Hypothetical):** Proposed by Stephen Hawking, this theoretical concept suggests that black holes can emit tiny amounts of radiation over time due to quantum effects near the event horizon. This radiation could cause black holes to slowly lose mass and eventually "evaporate."

- **Information Paradox (Hypothetical):** The question of what happens to information that falls into a black hole is a long-standing puzzle. According to quantum mechanics, information cannot be lost, but black holes seem to violate this principle. Resolving this paradox is a major challenge in physics.

- **Gravitational Time Dilation**: Near a black hole, time passes more slowly compared to a distant observer due to the extreme gravitational field. This phenomenon is known as gravitational time dilation.

- **Accretion Disks**: Matter drawn toward a black hole forms an accretion disk—a flattened, spiraling structure of gas and dust that heats up and emits high-energy radiation, including X-rays.

- **Black Hole Candidates**: Some astronomical objects, like Cygnus X-1 and Sagittarius A*, are strong candidates for containing black holes based on their observed behavior.

- **Black Hole Mergers (Hypothetical):** When two black holes orbit each other and eventually merge, they release powerful gravitational waves. These mergers have been detected through observatories like LIGO and Virgo.

- **Galactic Influence**: Supermassive black holes at the centers of galaxies can significantly influence the motion of stars and the evolution of galaxies themselves.

Structure of a Black Hole

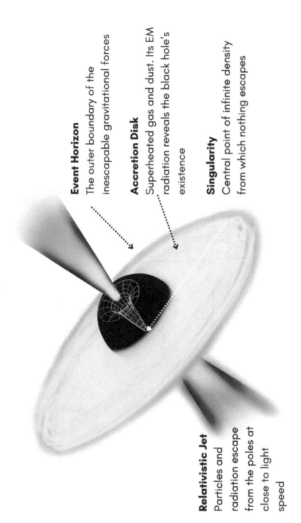

Event Horizon
The outer boundary of the inescapable gravitational forces

Accretion Disk
Superheated gas and dust. Its EM radiation reveals the black hole's existence

Singularity
Central point of infinite density from which nothing escapes

Relativistic Jet
Particles and radiation escape from the poles at close to light speed

Notes

Notes

Dark Matter

Dark matter remains one of the most significant unsolved mysteries in astrophysics and cosmology, and its nature continues to be a subject of intense scientific investigation.

- **Invisible and Undetectable**: Dark matter is a mysterious substance that cannot be observed directly with electromagnetic radiation because it neither emits nor absorbs light or other forms of electromagnetic radiation.

- **Gravitational Influence**: Dark matter is primarily detected through its gravitational effects on visible matter. It exerts a gravitational force that affects the motion of galaxies and galaxy clusters.

- **Abundance**: Dark matter makes up a significant portion of the universe's total mass and energy content. It is estimated to account for approximately 27% of the universe's composition.

- **Non-Baryonic Matter**: Dark matter is believed to consist of non-baryonic matter, which means it is not composed of the same particles as ordinary matter (protons, neutrons, and electrons).

- **Galaxy Rotation Curves**: Dark matter was first proposed to explain the observed flat rotation curves of galaxies, where the outer regions of galaxies rotate at higher speeds than expected based on visible matter alone.

- **Clustering and Structure Formation**: Dark matter plays an important role in the formation of cosmic structures, such as galaxies, galaxy clusters, and large-scale filaments. Its gravitational attraction helps gather and hold matter together.

- **Galactic Halos**: Dark matter is thought to exist in halos surrounding galaxies. These halos extend far beyond the visible boundaries of galaxies and contain a substantial amount of dark matter.

- **Missing Mass Problem**: Dark matter helps resolve the "missing mass problem" in the universe. Without dark matter, the observed gravitational effects would not match the visible matter's gravitational influence.

- **Cosmic Microwave Background**: Observations of the cosmic microwave background radiation provide strong evidence for the existence of dark matter. It affects the distribution of matter in the early universe.

- **Influence on Galaxy Formation**: Dark matter halos provide the gravitational scaffolding for the formation and evolution of galaxies. The distribution of dark matter affects how galaxies form and merge.

Interstellar Media (ISM)

Understanding the interstellar medium is essential for studying the life cycle of stars, galaxy formation, and the overall structure and dynamics of galaxies. It is not empty space but consists of gas and dust particles.

Interstellar media includes:

- Interstellar gas, mostly hydrogen and helium
- Interstellar dust, which is tiny solid particles, mostly carbon and silicates

Interstellar matter exists in various states, including atomic hydrogen (H I), molecular hydrogen (H2), which is important for star formation, and ionized gas (H II), which can be ionized by intense radiation from nearby stars.

Things SF writers may wish to incorporate into their worldbuilding:

- Interstellar dust particles scatter and absorb visible and ultraviolet light, making them visible as dark patches against bright background stars; they also help form protoplanetary disks.
- ISM is important for star formation.
- ISM contains magnetic fields.

- ISM interacts with cosmic rays which come from high-energy particles, ie. Supernovae.

- ISM can contain molecules, including complex organic molecules, as well as water (H_2O), carbon monoxide (CO), and formaldehyde (H_2CO).

ISM has been explored as a medium for potential **interstellar communication** through methods like the transmission of radio signals.

Introduction to Celestial Bodies

Celestial bodies are fascinating objects that make up the universe, and they help us explore the cosmos and understand our place in it.

- **Stars**: Stars are huge, hot balls of gas that emit light and heat through nuclear fusion. Our Sun is a star.

- **Planets**: Planets are large, solid or gas-covered objects that orbit stars. Earth, Mars, and Jupiter are examples of planets. One classification of a planet is that it has cleared its orbital neighborhood of other debris.

- **Moons**: Moons are natural satellites that orbit planets. Earth's moon is a familiar example.

- **Asteroids**: Asteroids are small, rocky objects that orbit the Sun, mostly found in the asteroid belt between Mars and Jupiter.

- **Comets**: Comets are icy objects that orbit the Sun and develop tails when they get close to it.

- **Galaxies**: Galaxies are vast collections of stars, gas, dust, and dark matter held together by gravity. The Milky Way is our galaxy.

- **Nebulas**: Nebulas are clouds of gas and dust where stars are born and can take beautiful shapes in space.

- **Black Holes**: Black holes are extremely dense regions where gravity is so strong that nothing can escape, not even light.

- **Exoplanets**: These are planets that orbit stars outside our solar system and are a focus of the search for extraterrestrial life.

- **Runaway Planets**: Sometimes a planet may be ejected from its solar system and travel across space until it is trapped in another gravity well or collides with something. These planets are cold, as they have no sun to warm their surfaces; however, they could be geothermally active and produce warmth from within.

- **Centaurs**: small bodies that have characteristics of both asteroids and comets, mostly found between Jupiter and Neptune. Chiron is an example.

- **Dwarf Planet**: is in direct orbit around the sun, smaller than any of the eight classical planets, and has cleared the neighborhood around its orbit. Ceres and Pluto are considered dwarf planets. Sometimes called a **planetoid**.

- **Kuiper Belt Objects (KBOs)**: small, icy objects and dwarf planets beyond Neptune.

- **Oort Cloud**: a hypothetical region of space, such as surrounding a solar system, filled with icy objects that are the source of comets. The Oort Cloud has not been scientifically observed.

Star Types

Main Sequence stars are classified by color, temperature, and magnitude.

Main Sequence Stars

Type	Color	Temperature	Magnitude	Example
O	Blue	> 30,000 °C	Brightest -5 to -1	10 Lacerta
B	Blue-white	10,000-30,000 °C	-4 to 0	Rigel
A	White	7,500-10,000 °C	0 to 2	Sirius
F	Yellow-white	6,000-7,500 °C	2 to 4	Procyon
G	Yellow	5,200-6,000 °C	4-ish	Sun
K	Orange	3,700-5,200 °C	4 to 6	Arcturus
M	Red	< 3,700 °C	Dimmest 6 to 15	Betelgeuse

Magnitude is "brightness" and the scale is backward: higher number = lower luminosity.

M class stars are the most common, about 80% of all stars. **G** stars are about 3.5%.

There are also:

- **Giant Stars**: Low mass stars (G, K, M) near the end of their lives

- **White Dwarfs**: Dying remnant of an imploded star (also classified as **D**)

- **Supergiant Stars**: High mass stars (O, B, A, F, G, K, or M) near the end of their lives

When making a world, consider the distance from and temperature of its star(s), because this affects its surface temperature and therefore the potential for liquid water. This is called the "Goldilocks Zone."

Stars often cluster in **globular clusters**. These clusters may contain millions of stars and are thought to be held together by gravity. Stars in globular clusters are thought to be the oldest in the universe.

The process of formation is not known, but it is believed it may be related to the collapse of dense gas clouds. The stars may move in orbits within the cluster (**dynamic equilibrium**).

Nebulae & Star Birth & Death

A nebula is a vast cloud of gas and dust in space. It can be incredibly large and spread out or relatively compact. Understanding the life cycle of stars and the role of nebulae is essential for comprehending the evolution of the universe, from the birth of stars to the creation of elements and the formation of planetary systems.

- A nebula forms from the remnants of previous stars (star explosions, or supernovae).

Star Birth

- New stars form within the nebula from the gas and dust in space (+gravity), when a region of a nebula becomes denser and cooler.

- A protostar will form in a dense, cool region where gas and dust clumps together as a hot, condensed core.

- When the core of a protostar becomes hot and dense enough (about 15 million degrees Celsius), nuclear fusion begins. Hydrogen atoms combine to form helium, releasing energy.

Star Death

- The fate of a star depends on its mass.

- Low-mass stars exhaust their hydrogen fuel and expand into red giants.

- Intermediate-mass stars can become more massive white dwarfs or neutron stars.

- Massive stars burn through their fuel quickly and end in a supernova explosion. Its core can collapse into a neutron star, or, rarely, become a black hole.

- Elements forged in stars are released into space through supernova explosions and become part of new nebulae, which can recycle into the start of future stars and planets.

Lifecycle of a Star

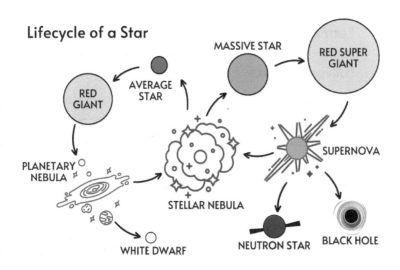

Cosmic Phenomena

Globular clusters and runaway stars are examples of intriguing cosmic phenomena that continue to captivate astronomers and contribute to our understanding of the universe's complexity and evolution. An example is Zeta Ophiuchi.

Globular clusters are densely packed, spherical groups of hundreds of thousands to millions of stars. They are some of the oldest known objects in the universe. Globular clusters are not galaxies but are often found in the halo regions of galaxies.

- **Composition**: These clusters primarily contain old stars, often with a high proportion of low-mass stars, as well as some more massive stars.

- **Formation**: Globular clusters are thought to have formed early in the universe's history from the collapse of massive molecular clouds. (+gravity)

- **Structure**: The stars in globular clusters are densely packed, with interactions between stars occurring more frequently than in the sparser regions of galaxies.

- **Orbital Motion**: Globular clusters orbit the centers of galaxies, including our Milky Way. They often follow highly elliptical orbits within the galactic halo, making periodic passes through the galactic plane.

- **Age**: Globular clusters are some of the oldest objects in the universe, with ages typically exceeding 10 billion years. Their stars formed when the universe was relatively young. This gives scientists a look at the early universe.

Runaway stars are individual stars that have been expelled from their original stellar environments, often traveling at high velocities through space.

- **Formation**: Runaway stars can be created through various processes:

  o Binary Interactions: A runaway star can form when one star in a binary system explodes in a supernova, ejecting the other star at high speed.

  o Stellar Encounters: Close gravitational encounters with other stars or massive objects within a star cluster can disrupt the orbit of a star, causing ejection.

- **Velocity**: Runaway stars can have velocities significantly higher than typical stars in their surroundings, often exceeding 100 kilometers per second (km/s).

- **Impact on Star Formation**: Runaway stars can influence the environments they travel through. They might trigger new star formation when their powerful stellar winds interact with nearby gas clouds.

Notes

Notes

Solar Systems

A solar system is a collection of celestial bodies orbiting or surrounding a star or a pair of stars, which provide light, heat, and gravitational pull that governs the orbits of objects in its system. Different solar systems may exhibit variations and unique characteristics, but these concepts apply broadly.

Formation and Evolution:

- Solar systems form from rotating clouds of gas and dust through processes like accretion and condensation.

- Their evolution may include the development of life and changes in planetary atmospheres.

Asteroids and Comets:

- Asteroids are rocky or metallic objects that orbit the central star, often found in the asteroid belt between Mars and Jupiter.

- Comets are icy bodies with highly elliptical orbits, known for developing bright tails when they approach the central star.

Solar Wind:

- The central star, like our Sun, emits a constant stream of charged particles known as the solar wind, which affects objects in the solar system.

Interaction with the Central Star:

- The central star's energy, including light and heat, influences the temperatures, climates, and conditions on planets and objects within the solar system.

Gravity:

- Gravity is the force that keeps objects in orbit around the central star.

- It depends on the mass of the central star and the distance from it.

Solar System Dynamics:

- The study of celestial mechanics and gravitational interactions helps predict the positions and movements of objects within the solar system.

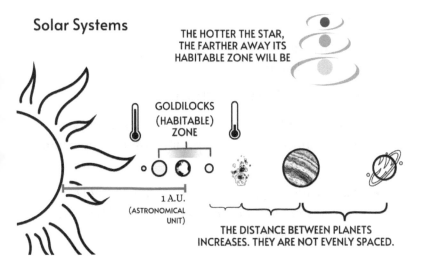

Solar Systems

THE HOTTER THE STAR, THE FARTHER AWAY ITS HABITABLE ZONE WILL BE

GOLDILOCKS (HABITABLE) ZONE

1 A.U. (ASTRONOMICAL UNIT)

THE DISTANCE BETWEEN PLANETS INCREASES. THEY ARE NOT EVENLY SPACED.

Planets & Worlds

Planets orbit stars, unless they have been ejected from their solar systems by some celestial force, such as a supernova. Many bodies that orbit larger planets, such as the moons of Jupiter, can be considered "worlds" as well. In general, a "world" is a solid body on which life might exist (naturally or invaded).

Planets:

- Planets are large celestial bodies that orbit the central star.

- They have spherical shapes and may have atmospheres and natural satellites (moons).

Moons (Natural Satellites):

- Many planets have one or more moons that orbit them.

- Moons are smaller bodies held in gravitational orbit around planets.

Terrestrial and Gas Giant Planets:

- Terrestrial planets (e.g., Earth, Mars) are rocky, have solid surfaces, and are typically closer to the central star.

- Gas giants (e.g., Jupiter, Saturn) are primarily composed of hydrogen and helium and have no (known) solid surfaces.

Ecliptic Plane:

- Most planets and objects in the solar system orbit in roughly the same plane called the ecliptic plane.

Orbitals:

- Planets follow elliptical (oval-shaped) orbital path around the central star. The star (or one of the stars) is usually at or near one of the ellipsis foci.

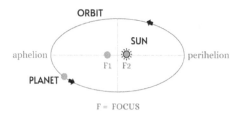

- The shape, size, and tilt of these orbits vary among planets.

- Each planet and moon has a specific orbital period, indicating the time it takes to complete one orbit around the central star or planet. Earth is 365.25 days or 1 year, for example. Mars takes 687 days, or about 2 years. A longer orbit means longer seasons on the planet.

- Given an elliptical orbit, as a planet nears its star it will speed up, and as it swings around it will then slow down. The farther a planet is from its star, the slower it will travel, but gravity will keep it in orbit. (Kepler's Laws)

Rotation and Tilt:

- Planets and celestial bodies rotate on their axes, determining day and night cycles. The direction and speed of rotation can vary.

- Planets and celestial bodies can have tilted axes. That means their north and south poles may not be perpendicular to the ecliptic plane. The tilted axes can cause seasonality.

- Earth is tilted about 23° off the perpendicular, which is why our daylight hours and the path of the sun in the sky changes during the year.

Uniformity:

- A planet's surface is uniform or lacks geographic variety. (Mars, Tatooine, Enceladus)

- A planet's surface can be highly geologically varied. (Earth)

- Different substances give planets different hues. [Mars = iron oxide (rust) = red]

- Cultures on different parts of a planet will adapt according to their local environment. Even if it is uniform, they will develop differently in different places. However, geography is a significant player in the development of cultures as well as biodiversity.

How Tilt Makes Seasons

SUMMER IN EUROPE

Note how the Northern hemisphere gets more sunlight

23° TILT

SUN

Poles get 24/7 light or 24/7 darkness

SUMMER IN AUSTRALIA

Note how the Southern hemisphere gets more sunlight

No Tilt = No Seasons or seasonal by latitude

Extreme Tilt = Extreme Seasons

Waves

Waves affect everything from galaxies to environments.

- **Mechanical waves**, which require a medium (substance) to travel through, such as sound waves (through air) and seismic waves (through the ground).

  o Because sound requires a medium, there is no sound in the vacuum of space.

- **Electromagnetic waves**, which do not require a medium, can propagate through a vacuum. Examples are x-rays, radio waves, and light waves.

  o In a vacuum, waves travel at a constant speed ($\sim 3 \times 10^8$ m/s, or light speed), but it slows in media.

  o EM waves are polarized to the orientation of the electric and magnetic fields that produce them.

Properties of waves:

- **Amplitude**: the hills and valleys – the distance from equilibrium to its maxiumum crest or trough. (m)

- **Frequency**: how many waves happen in a certain amount of time – such as 5 waves per second. (Hz)

- **Wavespeed**: how fast the wave travels. (m/s)

- **Wavelength**: the distance between troughs. (m)

- **Propagation**: how the wave moves - could be up-down, spirals, or back-and-forth. All waves propogate out spherically from the source.

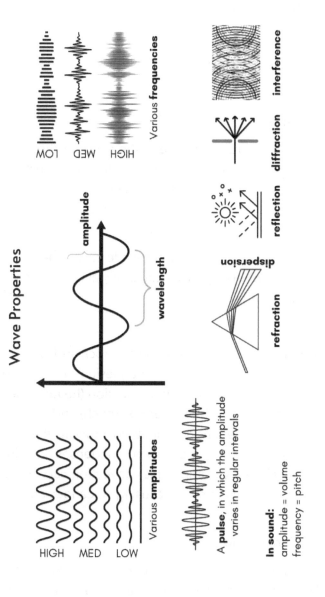

Various **frequencies**

LOW MED HIGH

interference

diffraction

reflection

dispersion

refraction

Wave Properties

amplitude

wavelength

Various **amplitudes**

HIGH MED LOW

A **pulse**, in which the amplitude varies in regular intervals

In sound:
amplitude = volume
frequency = pitch

Planet Layers

Every planet will have different conditions, construction, and geology, but Earth's construction is a good place to start.

Lithosphere

- The crust or rocky surface of the earth, or the "ground," including under water (oceanic crust). The lithosphere is made up of layers of solid rock. Under the rock, there is a layer of the upper mantle that behaves as a brittle, rigid solid.

- The partially molten upper mantle called the **asthenosphere** behaves plastically and is a dense, viscous fluid.

Mantle

- The mantle is made of semi-solid rock and is very hot. It is made of iron- and magnesium-rich silicate minerals. Its heat flows in conduction (heating the underside of the lithosphere by touch) and convection (heat rising from below toward the lithosphere, where it cools and condenses and sinks again, like a lava lamp). This convection causes the tectonic plates to move, causing volcanoes and earthquakes.

Outer Core (liquid)

- Primarily composed of liquid iron and nickel. It is liquid due to the high temperatures and pressures at its depth.

Inner Core (solid)

- Located at Earth's center, the inner core consists of solid iron and nickel. It remains solid due to the tremendous pressure at its depth. The core is believed to spin, which gives the Earth its magnetosphere.

Related but not a layer of the Earth:

Magnetosphere

- The magnetosphere is the magnetic field enveloping the Earth, believed to be caused by the spinning of our iron-nickel core.

- A weak magnetosphere can lead to strong solar and cosmic radiation. Protecting the planet form solar wind and particles helps prevent the atmosphere from being stripped away.

- The aurora borealis effect is the ionization of solar particles colliding with the magnetosphere and our thermosphere.

Layers of the Earth

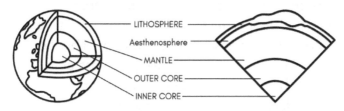

LITHOSPHERE

Aesthenosphere

MANTLE

OUTER CORE

INNER CORE

Gravity, Density, Mass

Density, mass, and size all play essential roles in determining the gravity of a planet.

The relationship between these three determines the gravity.

Mass: How many molecules are in a substance.

Density: How tightly packed the molecules of a substance are.

Size: When talking about planets (spheres), size would be radius or diameter.

Size and density determine mass. A planet with more mass will have more gravity.

Size matters because the farther you get from a planet's center, the gravitational force decreases by the inverse square law.

- Jupiter has size and mass, but because it is mostly gaseous, it has less density. If you collapsed or solidified Jupiter, its size would shrink, and it would become more dense, but its mass would be the same. The gravity would actually increase at its surface because it would be closer to the center; it would decrease relative to its moons due to the inverse square law.

- Suppose you have a solid sphere and a hollow sphere of the same size (golf ball and ping-pong ball). The solid one has more mass, therefore it would have more gravity.

- The moon has 1/6 the gravity of Earth. Its mass is 1/6 that of Earth. It is smaller, and probably less dense

This reference gives you an idea of different planet sizes and their respective surface gravity.

Comparison of World & Gravity

Celestial Body	Average Radius (kilometers)	Mass (kilograms)	Average Density (kg/m³)	Surface Gravity (m/s²)
Mars	3,389.50	6.42×10^{23}	3,933	3.72076
Earth	6,371	5.972×10^{24}	5,515	9.80665
Jupiter	69,911	1.898×10^{27}	1,326	24.79099
Pluto	1,188.30	1.303×10^{22}	1,850	0.622
Earth's Moon	1,737.1	7.342×10^{22}	3,344	1.625
Titan (Saturn's Moon)	2,575.5	1.345×10^{23}	1,882	1.352

Notes

Notes

Atmosphere

Atmospheres affect weather, climate, and the potential for life. The adaptations of living organisms evolving on a planet will depend on the atmosphere.

- Some planets have atmospheres **composed of gases** that envelop their surfaces.

- **Earth's moon** has a very thin atmosphere, mainly helium and argon, that extends about 10 km from the surface. Most life forms could not breathe such a thin atmosphere.

- **Venus has a very dense atmosphere**, which puts it at about 92 times the pressure of Earth's. It is mainly CO_2 (96.5%) and Nitrogen. Most life forms and machines would be crushed by this pressure. (Earth's equivalent is about 6 miles under water.)

- Atmospheres can be affected by **geology** (Venus), **rotation, distance** from the star, availability of **water, gas composition, magnetic field** (magnetosphere) (Mars's is very weak), and many other factors.

- **Atmospheric pressure** (and density) **decreases** with height. Think of the thinner atmosphere at the top of Mt. Everest, where climbers need to bring Oxygen.

- Earth's **atmospheric water varies** based on precipitation, temperature, and humidity.

- Earth's **atmospheric composition** is 78% Nitrogen (N2), 21% Oxygen (O2), < 1% Argon (Ar), 0.04% Carbon dioxide (CO2), 1% water vapor (depending on location), and trace other elements such as Neon, Helium, Methane (CH4, 0.00017%), Krypton, Hydrogen, Nitrous oxide, Sulfur dioxide, Ozone, Ammonia, and so on.

  o SCUBA tanks are designed to regulate the N2-O2 proportions to ensure divers get the right amount of oxygen with depth-pressure changes.

  o A breathing device or EVA suit for humans in other atmospheres would need to follow the same basic rules, depending on atmospheric pressure and composition.

- **Wind and Jet Streams**: Planetary atmospheres often have high-speed winds and jet streams that can impact weather and geography.

- **Greenhouse effect** is the trapping of heat in the atmosphere caused by CO2. Venus has a runaway greenhouse effect, making its atmosphere very hot.

- **Albedo** is the effect of solar radiation (light, heat) reflecting off the surface of the Earth. This happens best at the poles and glaciers, as well as with clouds.

- **Solar and Cosmic Radiation**: Atmospheric composition can influence how much solar and cosmic radiation reaches the planet's surface, affecting the health of inhabitants.

- **Chemical Reactions**: Chemical reactions within an atmosphere can have effects, such as the creation or destruction of ozone layers, as by pollutants.

The atmosphere has **multiple layers**, each with its own composition, density, and temperature.

- **Troposphere**: Closest to the surface, this is the layer that we know and breathe every day. This extends to about 12km above sea level. This layer is where airplanes fly and clouds appear.

- **Stratosphere**: This is the next layer, extending to about 50km. The ozone layer is within it, at about 20-30km.

- **Mesosphere**: About 50-80km high. This is the layer where meteors burn up.

- **Thermosphere**: About 80-600km high. Has no defined upper boundary and transitions into the exosphere. Has extreme temperature differential from top to bottom. Absorbs the most solar radiation (ie UV and X-ray). Ionization is responsible for auroras. Where satellites and spacecraft (ISS) orbit.

- **Exosphere**: Where Earth's atmosphere merges with the solar wind, a stream of charged particles emitted by the Sun. Scatters and absorbs sunlight. Creates the geocorona.

Layers of Earth's Atmosphere

Challenges for Spacecraft: The extreme temperatures and ionization in the thermosphere pose challenges for spacecraft and satellites. They can experience rapid heating on the sunlit side and cooling on the dark side, necessitating specialized thermal protection and communication systems.

Fiery Descent: Spacecraft experience "burning" or intense heat during reentry into a planet's atmosphere. This can occur in the exosphere or thermosphere, depending on the angle and trajectory of reentry. This is why spacecraft need heat shields. The heating is caused by the compression of air in front of the spacecraft due to its high speed. As it descends, this heat is dissipated in the troposphere and stratosphere.

Exposure to the Vacuum of Space

Exposure is an extremely hazardous and life-threatening situation. Here is a list of the processes and effects of human exposure to space vacuum:

- **Sudden Decompression**: A vacuum has zero pressure. Going from human-tolerable pressure to zero causes a rapid loss of air pressure. The skin is elastic enough to withstand it (until it freezes), but see more below.

- **Ebullism**: At low pressures, fluids boil at lower temperatures. Fluids in the body begin to boil, causing tissue damage.

- **Swelling and Bruising**: The low pressure auses body fluids, including saliva and the fluids in the eyes, to boil away, leading to swelling and bruising of exposed skin.

- **Severe Tissue Damage**: Due to the cold, low pressure, vaporization, and other factors, space can lead to severe tissue damage, particularly in the respiratory system, eyes, and skin. Think frostbite. Also, Unfiltered sunlight in space can lead to thermal burns on exposed skin.

- **Lack of Air**: There is no atmosphere to breathe. Within seconds, a person would lose consciousness due to hypoxia. A person might be tempted to hold their breath, but this will cause their lungs to expand, leading to lung barotrauma.

- **Radiation: Space** is filled with harmful ionizing radiation from the sun and cosmic rays, which can damage cells and DNA.

- **Unconsciousness and Death**: As oxygen levels drop, and possibly as blood to the brain changes pressure or boils, a person would become unconscious, then develop brain damage, and within a few minutes (likely sooner) would die.

- **Survivors**: Brief exposures to a vacuum often cause various health complications, including lung and cardiovascular issue.

- **Precautions and Space Suits**: Astronauts and cosmonauts wear specially designed space suits to protect against the vacuum of space, providing life support systems, thermal control, and radiation shielding.

Other Hazards:

- **High velocity particles** and objects such as space trash

- **Extreme cold and extreme heat.** Sun exposure will create deadly heat, which contrasts with the extreme cold on the "shady" side. This requires special suits.

- **Radiation**, in general, is prevalent in space. Spacecraft must have shielding to protect occupants. Some radiation passes through all objects.

Space-Time

+ General Relativity and Time Dilation

Space-Time:

- Space-time is the four-dimensional framework that combines space and time into a single entity.

- It's the stage upon which all events in the universe occur, according to Albert Einstein's theory of relativity.

- Some scientists speculate that quantum entanglement bypasses space-time when it comes to travel / teleportation.

General Relativity:

- General relativity is Einstein's theory that describes how gravity works in the universe.

- It posits that gravity, in the form of massive objects, like planets and stars, warp or curve space-time around them, causing objects to move along curved paths.

Time Dilation:

- Time dilation is the effect where time passes differently for observers in different gravitational fields or relative motion.

- It means that time can appear to move slower or faster depending on your position and motion in the universe.

Travel:

- When a craft approaches the speed of light, time "inside" the craft is slower than outside the craft. For example, an astronaut at FTL may experience a few minutes while hundreds of years pass on Earth (mathematically inaccurate example).

- When a craft moves at "celestial" speeds (as fast as celestial bodies but not close to the speed of light), time slows but the difference between the astronaut and the Earth-bound people will be less pronounced, possibly negligible.

- Example: A rocket goes to Pluto at an average of 30kps (Earth's average orbital speed). It would take about 6.25 years to get there. 30kps is significantly slower than 300,000kps, so time dilation would not have a significant effect. Length contraction and mass increase would also be negligible.

Travel Concepts

Science fiction often explores various hypothetical travel mechanisms for both faster-than-light (FTL) and "normal speed" space travel. Here's a list of some common concepts and how they work:

Faster-Than-Light (FTL) Travel Mechanisms:

Wormholes:

- Concept: Wormholes are hypothetical tunnels or shortcuts through space-time, connecting distant regions of the universe.

- How They Work: Wormholes are like cosmic tunnels that allow spacecraft to enter at one end and exit at another, effectively bypassing the vast distances between them.

Alcubierre Drive (Warp Drive):

- Concept: Inspired by general relativity, this drive contracts space in front of a spacecraft and expands it behind, creating a "warp bubble."

- How It Works: By warping space-time, the Alcubierre drive enables a ship to "ride" a wave of compressed space, theoretically allowing FTL travel.

Hyperspace:

- Concept: Hyperspace is an alternate dimension or region of space where the laws of physics differ, enabling faster-than-light travel.

- How It Works: Ships enter hyperspace, where distances are shorter or time behaves differently, facilitating faster travel.

Warp Gates/Stargates/Portals:

- Concept: These are artificial or natural portals that instantaneously transport objects from one gate to another.

- How They Work: When activated, a ship enters a gate and exits at another location, eliminating the need for conventional travel.

Quantum Entanglement ("Jump"):

- Concept: Quantum entanglement connects any two (or more) places. An advanced technology could harness entanglement to "jump" or teleport.

- How It Works: Information about the quantum state of one particle is transmitted to another particle at a distant location, effectively recreating the state of the first particle in the second particle.

"Normal Speed" Space Travel Mechanisms:

Chemical Rockets: Traditional rockets use chemical reactions for propulsion.

Ion Propulsion:

- Concept: Ion thrusters use electricity to accelerate ions as a propellant, achieving higher speeds than chemical rockets.

- How They Work: By ejecting ions at high velocities, ion thrusters provide efficient but gradual acceleration.

Solar Sails:

- Concept: Solar sails use pressure from photons in sunlight to propel spacecraft.

- How They Work: Thin, reflective sails capture and reflect photons, gradually accelerating the spacecraft over time.

Nuclear Propulsion:

- Concept: Nuclear engines harness the energy from nuclear reactions for thrust.

- How They Work: Controlled nuclear reactions heat propellant to create thrust, offering greater speed and efficiency than chemical rockets.

Fusion Propulsion (Hypothetical):

- Concept: Fusion drives would use controlled nuclear fusion as a power source for propulsion.

- How It Might Work: Fusion would generate immense energy for high-speed travel, but practical fusion drives remain theoretical.

Antimatter Propulsion (Hypothetical):

- Concept: Antimatter engines use antimatter as a fuel source to achieve near-light-speed travel.

- How It Might Work: When antimatter meets matter, it annihilates, releasing energy that can be used for propulsion.

Ramship (Ramjet Ship) (Hypothetical):

- Concept: Takes in particles, such as interstellar hydrogen, from surroundings, compresses it, and mixes it with fuel to create thrust.

- How it Might Work: Collected hydrogen is compressed to increase its density, becoming a propellant. This could be achieved using magnetic fields, nuclear reactors, or other energy sources.

Notes

Notes

Astrobiology

Understanding these key concepts in astrobiology can enhance the scientific realism and creativity of science fiction stories that explore the possibilities of life beyond Earth and the challenges of space exploration.

- **Habitability**: Habitability refers to the conditions required for a planet or celestial body to support life. Factors include liquid water, a stable and suitable environment, and the availability of essential elements and nutrients.

- **Exoplanets**: Exoplanets are planets that orbit stars outside our solar system. Astrobiology explores the potential habitability of exoplanets and the search for Earth-like planets in the habitable zone, where liquid water could exist

- **Extremophiles**: Extremophiles are microorganisms capable of surviving in extreme environments, such as high radiation, extreme temperatures, acidic conditions, or deep in the Earth's crust. Studying them helps scientists understand the limits of life on Earth and the potential for life on other planets. (Tardigrades?)

- **Mars Exploration**: Mars is a primary target for astrobiology research due to its potential habitability in the past and the search for signs of past or present microbial life. Science fiction often explores the idea of Martian life.

- **Europa and Enceladus**: Moons like Europa (around Jupiter) and Enceladus (around Saturn) have subsurface oceans beneath icy crusts. These environments are of astrobiological interest due to the possibility of liquid water and potential habitability.

- **Fermi Paradox**: The Fermi Paradox refers to the apparent contradiction between the high probability of extraterrestrial life in the universe and the lack of evidence or contact with advanced extraterrestrial civilizations.

- **Panspermia**: Panspermia is the theory that life, or the building blocks of life, can be transferred between celestial bodies, potentially allowing life to spread throughout the universe.

- **Extreme Environments on Earth**: Many science fiction scenarios involve life thriving in extreme environments on Earth, such as deep-sea hydrothermal vents, acid mine drainage, or salt flats. These environments inform speculation about life on other planets.

- **Biosecurity and Planetary Protection**: Astrobiology also addresses ethical and practical concerns related to space exploration, such as the risk of contaminating other celestial bodies with terrestrial life.

Definition of Life

Biologists generally agree on a set of characteristics that distinguish living organisms from non-living matter. These characteristics, often referred to as the "characteristics of life" or "properties of life," provide a framework for understanding and identifying living entities.

- **Cellular Structure**: Living organisms are composed of one or more cells, which are the fundamental units of life. Cells can be simple, as in bacteria, or complex, as in multicellular organisms like plants and animals.

- **Organization**: Life exhibits a high degree of organization. Organelles are organized into cells, which organize into tissues, organs, and organ systems in multicellular organisms. This hierarchical organization allows for specialized functions.

- **Energy Processing**: Living organisms require energy to carry out life processes. They obtain energy through various metabolic processes, such as photosynthesis (in plants) or cellular respiration (in animals).

- **Response to Stimuli**: Living organisms can detect and respond to changes in their environment. This includes responding to external stimuli like light, temperature, and chemicals, as well as internal signals.

- **Homeostasis**: Organisms maintain a stable internal environment, known as homeostasis, by regulating their internal conditions. This ensures that essential processes can function optimally.

- **Growth and Development**: Living organisms can grow in size and complexity over time. Growth involves an increase in cell number and size, while development refers to the process of maturation and specialization.

- **Reproduction**: Living organisms can reproduce, producing offspring that inherit genetic information from their parents. Reproduction can occur through asexual or sexual means.

- **Adaptation**: Living organisms can adapt to their environment over time through evolutionary processes. This adaptation involves changes in the genetic makeup of populations, leading to improved survival and reproduction.

- **Genetic Material**: Living organisms store genetic information as DNA. This information encodes the instructions for the structure and function of the organism.

- **Evolutionary History**: All living organisms in a biosphere share a common evolutionary history and are part of the tree of life. They have evolved from common ancestors and continue to evolve over time.

Sentience & Sapience

What kind of life might we find on other planets, and what are the ethical concerns for such living organisms?

The terms "sentience" and "sapience" are often used to describe different levels of cognitive and conscious abilities in beings, particularly in the context of intelligence and self-awareness.

Pre-Sentience:

- Pre-sentience refers to the absence of conscious awareness or subjective experience. Beings at this level **do not possess self-awareness** or the capacity to perceive and interpret their surroundings in a conscious or sentient manner. They may exhibit basic reflexes and instinctual behaviors but lack true awareness of their actions or environment. (See definition of life above, response to stimuli.)

- Examples: bacteria, fungi, plants, fish?

How do we know whether an organism possesses self-awareness? This is a question many scientists and non-scientists ask.

Sentience:

- Sentience is **the capacity to perceive and experience the world subjectively**. Sentient beings are <u>conscious and aware of their surroundings</u>, can experience sensations like pleasure and pain, and <u>have the ability to make choices based on their perceptions</u>. Sentience often involves the ability to experience emotions and respond to stimuli <u>in a meaningful way</u>. Many animals, including mammals and birds, are considered sentient.

- Examples: mice, cockroaches, pigs

Sapience:

- Sapience, often referred to as **higher consciousness or rational thought**, represents a higher level of cognitive abilities. Beings with sapience possess not only sentience but also <u>the capacity for complex reasoning, abstract thinking, problem-solving, and self-awareness</u>. They can engage in conceptual thought, learn from experience, plan for the future, and possess a sense of morality and ethics. Human beings are typically regarded as sapient creatures.

- Examples: humans, Wookiees, dolphins?, squirrels*

If you knew squirrels like I do, you would not argue.

Ecosystems

Creating a planet with a diverse and vibrant ecosystem can be a fascinating endeavor.

- **Biodiversity**: A thriving ecosystem should have a wide variety of species, including plants, animals, fungi, and microorganisms. Consider inventing unique and exotic life forms adapted to the planet's specific conditions.

- **Food Web**: Ecosystems are characterized by complex food webs where organisms are interconnected through predator-prey relationships and mutual dependencies. Think about the roles different species play in the food chain.

- **Habitat Diversity**: Create diverse habitats, such as forests, grasslands, wetlands, deserts, mountains, and aquatic environments. Each habitat can support different species and ecological niches.

- **Climate Zones**: Design climate zones on your planet, including polar, temperate, tropical, and arid regions. Climate variations can influence the types of life forms that thrive in each zone.

- **Adaptations**: Consider the unique adaptations of organisms to the planet's environment. Think about how they have evolved to survive in extreme temperatures, high radiation, or unusual atmospheric conditions.

- **Keystone Species**: Introduce keystone species, which have a disproportionately large impact on their ecosystem. The removal of a keystone species can lead to significant ecological changes.

- **Symbiotic Relationships**: Explore symbiotic relationships between species, such as mutualism (mutually beneficial interactions), parasitism (one benefits at the expense of the other), and commensalism (one benefits without affecting the other).

- **Endemism**: Create endemic species that are unique to your planet. These species exist only on this world and contribute to its distinctiveness.

- **Ecological Succession**: Consider how ecosystems change over time through processes like primary succession (starting from bare rock or soil) and secondary succession (after disturbances like wildfires or floods).

- **Biogeochemical Cycles**: Develop cycles for essential elements like carbon, nitrogen, and water. These cycles influence the availability of nutrients for plant growth and other life processes.

- **Predator-Prey Dynamics**: Explore the interactions between predators and prey, including population fluctuations and adaptations for predation and defense.

- **Ecosystem Services**: Think about the services your ecosystem provides to its inhabitants, such as pollination, water purification, and nutrient cycling.

- **Ecological Resilience**: Consider how your ecosystem responds to disturbances, such as natural disasters or climate changes. Resilient ecosystems can recover from disruptions.

- **Cultural Significance**: Include elements of cultural significance, such as species revered or important to the indigenous inhabitants of your planet.

- **Environmental Challenges**: Incorporate environmental challenges or threats that impact the ecosystem, such as habitat loss, invasive species, or pollution.

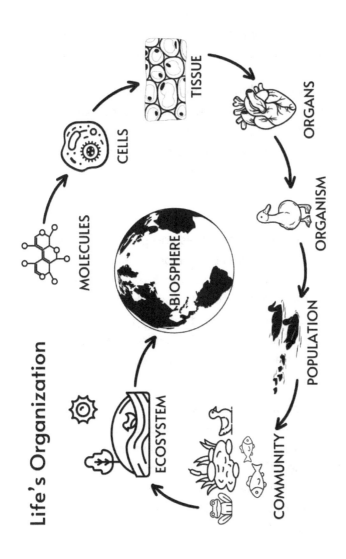

Life's Organization

Evolution

Alien species don't come out of nowhere. They must evolve from something, and the traits they inherit will be suited to survive in their environment/ecosystem.

- **Variation**: Within a population of organisms, there is genetic variation due to mutations, genetic recombination, and other factors. Variation is important within a species because it allows for more opportunities to survive when conditions change.

- **Selection**: Environmental pressures, such as competition for resources and predation, create selective pressures that favor certain traits over others. For example, in Earth's early history, oxygen was trapped in water. Photosynthetic microorganisms created the oxygen we breathe today. "Fish" that could breathe both in water and oxygen had a selective trait to evolve into land animals.

- **Adaptation**: Individuals with advantageous traits that enhance their survival and reproduction are more likely to pass those traits to their offspring. With the selection example above, those "land fish" had the advantage of avoiding predators while on land, or of finding mates in other water sources to reproduce.

- **Heritability**: Beneficial traits are often heritable, meaning they can be passed on to the next generation through genetic inheritance. There are dominant and recessive traits that create variety and affect an organism's adaptability.

- **Reproduction**: Organisms with advantageous traits tend to reproduce more successfully, producing more offspring than those with less advantageous traits.

- **Natural Selection**: Over time, the population shifts toward individuals with the advantageous traits through a process known as natural selection. Harmful traits *tend* to be weeded out when those traits prove to be less advantageous, or they might persist through the generations until certain conditions are met.

- **Speciation**: Accumulated changes in a population's genetic makeup can lead to the formation of new species when populations become reproductively isolated. For example, the different species of turtles living on different islands in the Galapagos.

- **Diversity**: Evolutionary processes, including mutation and genetic drift, contribute to the diversity of life on Earth, resulting in the multitude of species we observe today. Migration and isolation also create diversity.

DNA & Reproduction

DNA is the building blocks of life. Viruses can have either DNA or RNA, but viruses are not considered to be living organisms.

- **Note**: It is believed by some that DNA can spontaneously respond to environmental conditions, but such a mutation is not always passed on, and most of the time, DNA changes randomly during cell production (ie meiosis). Whether or not the resulting traits are advantageous depends on their expression and the environment.

- **Mitosis** is cell division. Skin cells, for example, make more skin cells by mitosis. Most often, the daughter cells are identical (clones).

- **Meiosis** is the cell division process only for the production of gametes, or sex cells. In meiosis, genes can recombine randomly, which creates variation in individuals and within populations.

- **Gametes** are sex cells, such as oocytes (eggs) and sperm cells.. They are usually half the DNA from one parent and half from the other, forming a **zygote**.

- **Sexual reproduction** is the combination of two (or more, depending on the species) gametes to form a new cell with both parents' copies of DNA.

- **Asexual reproduction** is the duplication of a cell into two daughter cells through mitosis.

Reproduction in Plants & Fungi

Flowering Plants

- Plant reproduction begins with a **flower**.

- Inside, the **pistil** contains the ova and the **stigma**, which sticks out of the flower, produces **pollen**.

- When pollen lands on a flower and travels down to the pistil to fertilize the **ovules** (eggs) in the ovary.

- Fertilized eggs become **seeds**, which grow inside the ovary until ready to be released.

- The flower falls away and the ovary grows into a fruit.

Fungi

- Fungi asexually produce gamete-like cells in the form of **spores**.

- Spores are released into the environment when conditions are right. They can be carried on wind, water, or even on an animal's fur.

- Spores that land in a suitable place will start growing into new fungi.

- The spore sends out tiny threads called **hyphae** that spread and form a network called **mycelium**, which absorbs nutrients and is usually unseen.

- When the mycelium produces its special structures, like **mushroom** caps, it releases more spores.

Notes

Notes

Big Thank You

I would like to thank my husband for letting me have the time to create this guidebook. As a former science teacher, I am always excited to share scientific knowledge. If you find anything glaringly inaccurate, please let me know via email (see my web site).

Mostly I thank YOU for either buying this book, or for signing up for my newsletter, *Why Bards*.

If you haven't, **you can sign up directly** at https://whybards.substack.com.

It's free, and you can always unsubscribe at any time (no hard feelings).

The newsletter is about creative endeavors, including art appreciation, interviews with various types of artists (musicians, filmmakers, authors, painters…), and some occasional rambling articles about creativity-related stuff.

It will also be where I announce new books and my novels.

♥ PLEASE LEAVE A REVIEW AND SHARE THE LOVE! ♥

About the Author

Ingrid Moon is a lifelong science and pop culture nerd and animal lover. She has a bunch of useless degrees and certifications, one of which is a science teaching credential. She has sent experiments to the ISS and climbed high mountains. She currently lives in Los Angeles with her husband, teenager, six cats, two pigs, and a plethora of squirrels.

https://ingridmoon.com

Made in United States
Troutdale, OR
12/12/2023

15774388R00046